ANIMAL BATTLES

KOMODO DRAGON VS. ORANGUTAN

TORQUE™

BY NATHAN SOMMER

BELLWETHER MEDIA • MINNEAPOLIS, MN

Torque brims with excitement perfect for thrill-seekers of all kinds. Discover daring survival skills, explore uncharted worlds, and marvel at mighty engines and extreme sports. In *Torque* books, anything can happen. Are you ready?

This edition first published in 2021 by Bellwether Media, Inc.

Library of Congress Cataloging-in-Publication Data

Names: Sommer, Nathan, author.
Title: Komodo dragon vs. orangutan / Nathan Sommer.
Other titles: Animal battles (Bellwether Media)
Description: Minneapolis, MN : Bellwether Media, 2021. | Series: Torque. Animal battles | Includes bibliographical references and index. | Audience: Ages 7-12. | Audience: Grades 4-6. | Summary: "Amazing photography accompanies engaging information about the fighting capabilities of Komodo dragons and orangutans. The combination of high-interest subject matter and light text is intended for students in grades 3 through 7"–Provided by publisher.
Identifiers: LCCN 2020003030 (print) | LCCN 2020003031 (ebook) | ISBN 9781644872819 (library binding) | ISBN 9781681037448 (ebook)
Subjects: LCSH: Komodo dragon–Juvenile literature. | Orangutans–Juvenile literature.
Classification: LCC QL666.L29 S66 2021 (print) | LCC QL666.L29 (ebook) | DDC 597.95/968–dc23
LC record available at https://lccn.loc.gov/2020003030
LC ebook record available at https://lccn.loc.gov/2020003031

Editor: Kieran Downs Designer: Andrea Schneider

Printed in the United States of America, North Mankato, MN.

TABLE OF CONTENTS

THE COMPETITORS

The islands of Indonesia house beautiful scenery and a lot of wildlife. On some of these islands, the Komodo dragon rules. These massive lizards eat whatever **prey** they want.

But Komodo dragons have a challenger. Orangutans are some of the strongest and smartest **primates**. Who would win in a battle between these beasts?

Komodo dragons are the world's largest lizards. They grow as long as 10 feet (3 meters) and weigh more than 300 pounds (136 kilograms)! These **predators** have thick bodies with long heads and tails.

Komodo dragons only live on five Indonesian islands. They prefer **tropical** forests and grassy areas. But they also search for prey on beaches.

KOMODO DRAGON PROFILE

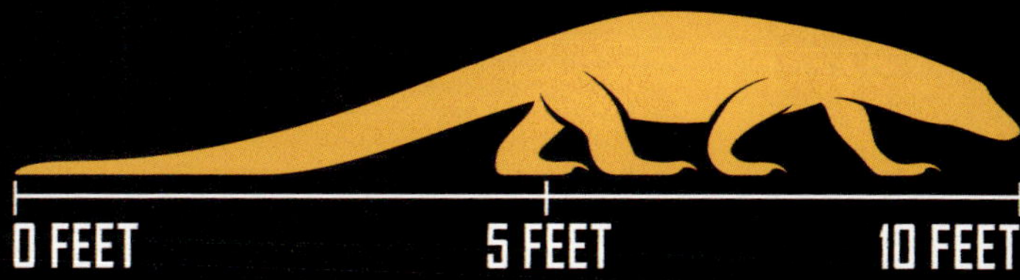

LENGTH
UP TO 10 FEET (3 METERS)

WEIGHT
MORE THAN 300 POUNDS (136 KILOGRAMS)

HABITAT

KOMODO DRAGON RANGE

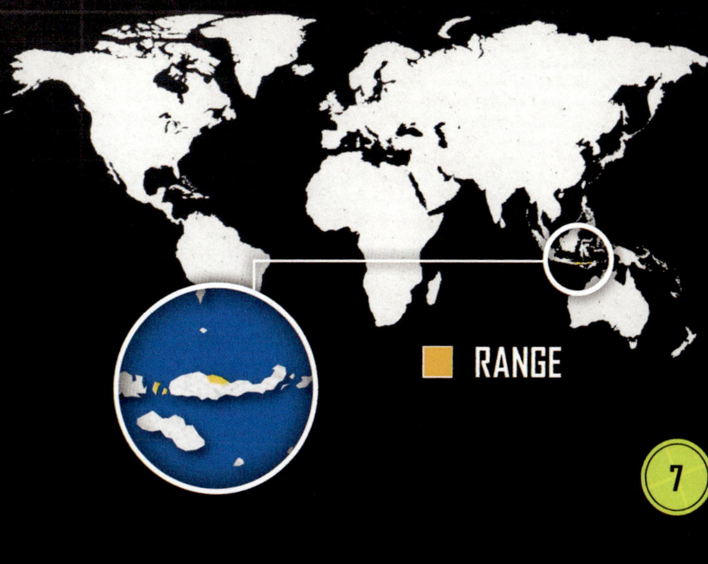

ORANGUTAN PROFILE

HEIGHT
UP TO 4.3 FEET
(1.3 METERS)

WEIGHT
UP TO 286 POUNDS
(130 KILOGRAMS)

HABITAT

RAIN FORESTS

ORANGUTAN RANGE

Orangutans are **apes** that stand up to 4.3 feet (1.3 meters) tall and weigh up to 286 pounds (130 kilograms). They have flat noses, sturdy bodies, and red fur.

Orangutans live in Indonesia and Malaysia. Most are found in **rain forests**. The primates spend much of their days high up in treetops.

SECRET WEAPONS

Orangutans can get around quickly. These **agile** beasts walk and swing through the **canopy**. Hooklike fingers and toes help them easily climb trees as an escape.

KOMODO DRAGON SPEED

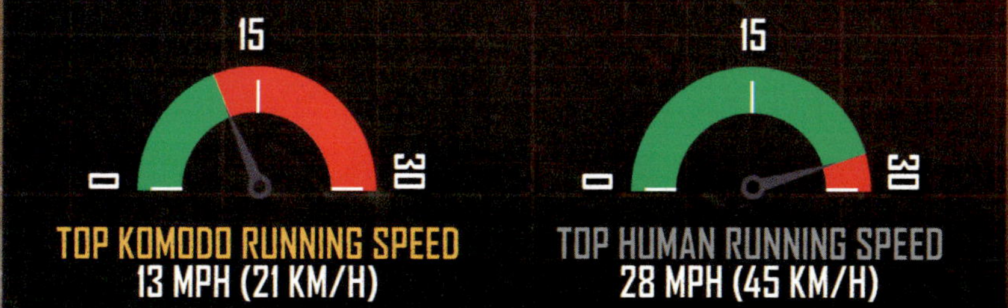

Powerful legs give Komodo dragons speed. The lizards can run up to 13 miles (21 kilometers) per hour in short bursts. They can easily catch slower prey.

SECRET WEAPONS

KOMODO DRAGON

SPEED

SHARP CLAWS

DEADLY VENOM

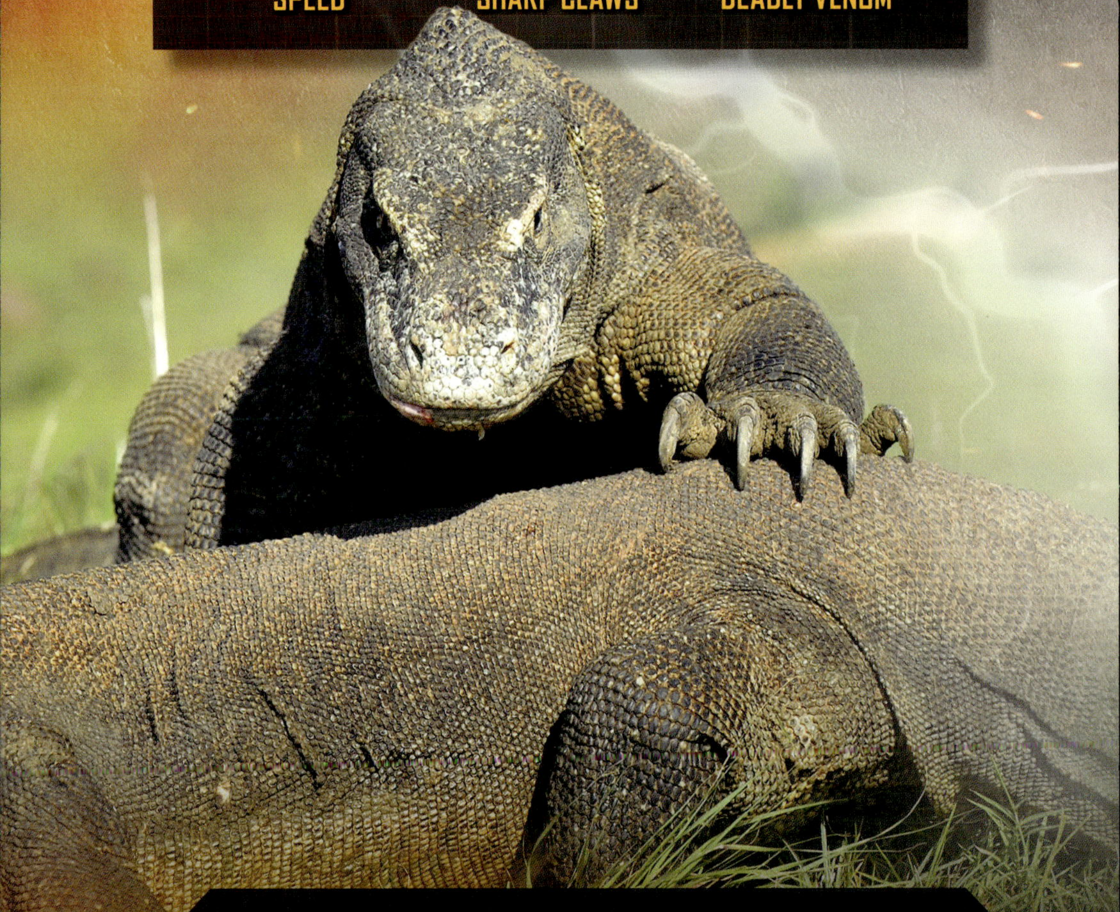

Komodo dragons have long, razor-sharp claws. These easily wound enemies. Their claws also help the lizards attack and hold their prey.

SECRET WEAPONS

ORANGUTAN

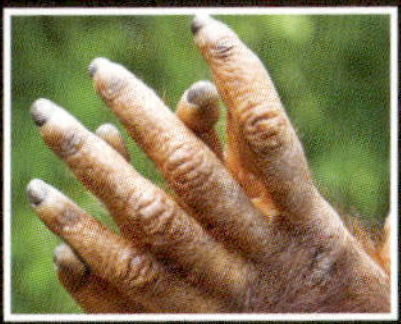

HOOKLIKE FINGERS

INTELLIGENCE

STRENGTH

Orangutans use **intelligence** to outsmart enemies. They warn each other when danger is spotted. Sometimes they wait to sound the alarm so they can sneak away quietly.

Komodo dragons have a deadly bite. They release **venom** that stops blood from flowing properly. This slowly kills prey.

NEXT-LEVEL SENSES

Forked tongues help Komodo dragons smell rotting prey from miles away. They can also see as far as 985 feet (300 meters).

ORANGUTAN STRENGTH

Orangutan arms get strong from swinging through trees. The apes are seven times stronger than humans! They can use this strength to overpower enemies.

ATTACK MOVES

Komodo dragons **stalk** and **ambush** enemies. The lizards spend hours waiting for prey to pass by. Then they attack at the perfect moment!

Orangutans do not fight often. They spend most of their time safe in the trees. There, they hide themselves and call out to other orangutans to warn of danger.

JUST LIKE US

Orangutans have a lot in common with humans. The word *orangutan* means "person of the forest" in the Malay language.

The bite of Komodo dragons usually gets the job done. Most enemies die from shock and blood loss. Komodo dragons follow wounded prey until it dies.

BIG APPETITE

Wide jaws and large stomachs help Komodo dragons eat a lot. They can eat more than half of their body weight in one meal.

Orangutans charge at attackers if forced to fight. They reach with their long arms to scratch foes. They also bite if enemies come close enough.

READY, FIGHT!

A Komodo dragon hides as an orangutan approaches. The lizard sneakily rushes and bites the ape as it walks by.

The orangutan tries to bite the Komodo dragon. But the lizard uses its speed to escape! The dragon's venom slowly poisons the orangutan. Komodo dragons rule the rain forest today!

GLOSSARY

agile—able to move quickly and easily

ambush—to carry out a surprise attack

apes—large primates that lack a tail

canopy—the uppermost level of the rain forest

intelligence—the ability to learn or understand

predators—animals that hunt other animals for food

prey—animals that are hunted by other animals for food

primates—any of a group of mammals that includes humans, apes, and monkeys

rain forests—thick, green forests that receive a lot of rain

stalk—to follow closely and quietly

tropical—having to do with a place that is hot and wet

venom—a poison made by Komodo dragons

TO LEARN MORE

AT THE LIBRARY

Grunbaum, Mara. *Orangutans.* New York, N.Y.: Children's Press, 2018.

Kingsley, Imogen. *Komodo Dragons.* Mankato, Minn.: Amicus, 2019.

Sommer, Nathan. *Gorilla vs. Leopard.* Minneapolis, Minn.: Bellwether Media, 2020.

ON THE WEB

FACTSURFER

Factsurfer.com gives you a safe, fun way to find more information.

1. Go to www.factsurfer.com
2. Enter "Komodo dragon vs. orangutan" into the search box and click 🔍.
3. Select your book cover to see a list of related content.

INDEX

The images in this book are reproduced through the courtesy of: Sergey Uryadnikov, front cover (Komodo dragon), pp., 11, 12, 14, 17, 18, 21 (Komodo dragon); Sermsak S, front cover (orangutan); GUDKOV ANDREY, pp. 4, 13 (strength), 19; vedderman123, p. 5; Danny Ye pp. 6-7; Donyanedomam, pp 8-9; Dan Olsen, p. 10; Lillian Tveit, p. 10 (hooklike fingers); Rhizma, p. 12 (speed); Adhi Rachdian, p. 12 (sharp claws); David Evison, p. 12 (deadly venom); meunierd, p. 13; undefined undefined, p. 13 (long fingers); Avalon/Photoshot License / Alamy Stock Photo, p. 13 (intelligence); USO, p. 15; Gameoper, p. 16; Charlotte Bleijenberg, pp., 20-21 (orangutan).